# The Hologram Education Institute

# The Hologram Education Institute

Mahendra Jagir

**Studio of Books LLC**
5900 Balcones Drive Suite 100
Austin, Texas 78731
*www.studioofbooks.org*
Hotline: (254) 800-1183

Ordering Information:
Special discounts are available on quantity purchases by corporations, associations, and others. For details, contact the publisher at the address above.

Printed in the United States of America.

ISBN-13:  Paperback    978-1-970283-62-4
          Hardback     978-1-970283-63-1
          eBook        978-1-970283-64-8

Library of Congress Control Number:    2026907495

# DEDICATION

This work, and the future it strives to protect and illuminate, is lovingly dedicated to my grandchildren:
**Dustin, Matthew, Georgia, and Jolie.**

You are not only the inspiration behind these pages, but the enduring reason this vision was created. Every idea, framework, and system described within this work is ultimately rooted in a single hope—that your generation will inherit a world that expands your possibilities rather than limiting them. In a future that may grow increasingly crowded, complex, and uncertain, may you always find space where it matters most: within your thoughts, your emotions, and your imagination.

May you never feel confined by the physical boundaries of the world around you. Instead, may you be guided by curiosity that refuses to diminish, courage that strengthens with time, and imagination that transforms limitation into opportunity. Let knowledge be your compass in moments of uncertainty, and let your heritage serve as your anchor—steadying you as you navigate the vast and shifting horizons of a changing world.

In this envisioned model of the future, you are no longer positioned as passive learners within a fixed system. You are recognized as **Guardians of the Earth**—individuals entrusted not only with understanding the world, but with sustaining and shaping it.

**Dustin & Matthew:** You are envisioned as the stewards of structure and systems, those who understand the language of mechanics, infrastructure, and energy. Through your mastery, you carry forward the principles embedded within SHIELD and SAN—ensuring that the frameworks of stability, connectivity, and resilience continue to function for generations beyond your own. Your role is to understand how systems breathe, how they endure, and how they evolve without breaking.

**Georgia & Jolie:** You are envisioned as the guardians of memory, life, and continuity. Through your care and insight, you preserve the wisdom contained within the Silent Seeds Archive and the depth of Fiji's long and layered history. Your role is to ensure that life is not only sustained, but remembered—that culture, ecology, and identity remain alive within the unfolding story of humanity.

Together, you represent more than individual futures—you form the four foundational pillars of a world designed not merely to endure its own expansion, but to thrive because of the balance you maintain within it. Each of you holds a distinct responsibility, yet all are interconnected within a shared system of purpose, stewardship, and vision.

You are not simply inheritors of what has been built—you are its continuation, its refinement, and its forward motion. Within you exists the capacity to shape a future that is more thoughtful, more compassionate, and more interconnected than the one you were given.

Stand firmly in your sovereignty. Learn without limitation. Create without hesitation.

We have already traced the foundations for you—the subterranean intelligence of SAN, the resilient energy systems of SHIELD, the framework of digital sovereignty, and the complete educational continuum of the Hologram Education Institute, spanning from early learning to doctoral mastery.

Now, what remains is not the construction of the path, but your willingness to walk it, extend it, and redefine it.

You are the architects of a new era—and the enduring light that will carry it forward.

# TABLE OF CONTENTS

# PREFACE

## THE ARCHITECT'S INTENT

The world of 2026 stands as a profound paradox—an era defined simultaneously by unprecedented connection and increasing constraint. Humanity has woven an intricate web of communication, technology, and shared knowledge that spans the globe, linking individuals, cultures, and systems in ways once unimaginable. Yet beneath this extraordinary connectivity lies a growing tension: the physical limits of a finite planet are becoming more visible, more pressing, and more unavoidable.

We are connected—but not equally free.

We are advanced—but not equally secure.

We are informed—but not equally empowered.

As populations expand, urban centers densify, and environmental pressures intensify, the systems that once supported human development are being stretched beyond their original design. Education, in particular, finds itself at a critical threshold. Built for a different era—one of slower growth, stable environments, and localized communities—it now struggles to meet the demands of a world defined by scale, speed, and complexity.

It is within this tension that this work begins.

### A Personal Origin Within a Global Challenge

This book did not emerge from abstraction or distant speculation. It was not conceived as an intellectual exercise or a purely technological ambition. It was born from a deeply personal realization—one that exists at the intersection of love, responsibility, and foresight.

I began to ask a simple but urgent question:

## What kind of world will my grandchildren inherit?

Dustin, Matthew, Georgia, and Jolie represent more than the next generation of my family—they represent the future of humanity itself. Their lives will unfold in a world far more complex, interconnected, and constrained than the one we have known. The challenges they will face will not be isolated or temporary; they will be systemic, continuous, and deeply intertwined with the structures we leave behind.

To prepare them for such a world requires more than preservation of what exists. It requires **transformation**.

Their future demands systems that do not merely sustain the present, but actively expand the possibilities of what can be achieved within limitation. It demands a rethinking of how knowledge is accessed, how opportunity is distributed, and how civilization maintains balance under pressure.

This work is my response to that responsibility.

## The Motivation Behind HEI: A Necessary Evolution

The Hologram Education Institute (HEI) was conceived from a critical recognition: that traditional educational systems, while foundational to human progress, are no longer sufficient for the realities of the modern world.

The physical classroom—once a symbol of opportunity—has become, in many cases, a point of limitation. It is bound by geography, dependent on infrastructure, and often inaccessible to those who need it most. Its reach is finite, its capacity constrained, and its resilience vulnerable to disruption.

In a world defined by digital interconnection, environmental uncertainty, and global imbalance, education must evolve beyond these constraints.

HEI is not an incremental improvement. It is a **structural reimagining**.

It proposes that learning should not be tied to place, privilege, or proximity, but should exist as a **continuous, immersive, and universally accessible condition of human life**. It envisions a system where knowledge is not delivered from a distance, but experienced directly—where every learner, regardless of location, can engage with education as an active, embodied reality.

At its core, HEI is driven by a simple but powerful belief:

**No child's potential should be limited by where they are born.**

### A Vision for Future Generations

The vision that guides this work extends far beyond the present moment. It is not designed for immediate convenience, but for long-term continuity. It is a vision of a world where future generations are not confined by the limitations of physical space, but empowered by systems that expand alongside their curiosity and imagination.

For Dustin, Matthew, Georgia, and Jolie, this vision represents something deeply personal:

A world where their environment does not restrict them, but responds to them.
A world where knowledge is not distant, but surrounding.
A world where learning is not imposed, but discovered.

In this future, education becomes infinite in reach, adaptive in structure, and intimate in experience. It is no longer something that is completed within a phase of life—it becomes a lifelong continuum, evolving as the individual evolves.

Knowledge is no longer inherited passively. It is actively shaped, questioned, expanded, and lived.

### The Convergence of Cultural Memory and Technical Design

This work also exists at a critical intersection—between **cultural preservation and technological innovation**.

As the world accelerates toward globalization and uniformity, there is a growing risk that cultural identities, ancestral knowledge, and historical narratives may be diluted or lost. This is particularly true for regions rich in oral tradition and ecological wisdom, such as Fiji.

The concept of Fiji's "Long Lost History" within this work is not simply about recovering the past—it is about ensuring that the past remains **alive within the future**.

Cultural memory is not treated as static record, but as a living system—one that must be experienced, understood, and carried forward. It provides grounding, identity, and meaning in a world that is otherwise rapidly shifting.

At the same time, this cultural foundation is paired with advanced technical systems such as the SHIELD energy infrastructure and broader civilizational frameworks. These systems ensure that the continuity of knowledge is supported by the continuity of energy, stability, and environmental balance.

Together, these elements form a dual foundation:

- **Cultural memory**, which defines who we are

- **Technical infrastructure**, which ensures we endure

Neither can exist meaningfully without the other.

### Three Interdependent Imperatives

At its deepest level, this work emerges from the convergence of three essential and interdependent needs:

1. **The preservation of cultural identity and historical continuity**
2. **The protection and stabilization of energy and environmental systems**
3. **The transformation of education into a borderless, resilient, and scalable framework**

These are not separate challenges. They are interconnected dimensions of the same fundamental question:

**How does civilization sustain itself across generations in a world of increasing complexity?**

## A Declaration of Intent

This preface stands as both a reflection and a declaration.

It reflects the reality of the world as it exists today—its limitations, its imbalances, and its vulnerabilities.

But it also declares a commitment to something greater: the deliberate design of systems that address these challenges not temporarily, but structurally.

The Hologram Education Institute (HEI) is not presented here as a finished solution, but as a **foundational framework**—one that can evolve, adapt, and grow with the needs of those who inherit it.

It represents a shift in thinking:

From scarcity —> to expansion

From limitation —> to possibility

From fragmentation —> to integration

## A Legacy Beyond Structure

Ultimately, this work is more than a proposal for a new educational system. It is a legacy—one shaped by the understanding that the future is not something we enter passively, but something we construct intentionally.

Through HEI, even as the physical surface of our planet becomes increasingly dense and constrained, the intellectual, creative, and cultural dimensions of humanity remain boundless.

It is a declaration that no matter how limited physical space may become, the space within the human mind will always remain infinite—if we build the systems to support it.

## Final Reflection

This is not simply a system.

It is a commitment—to continuity, to resilience, and to the generations who will carry the weight of what we create.

It is built for those who will inherit the world.
It is guided by those who choose to protect it.

And above all, it is a reminder that the future is not something we wait for—**it is something we are responsible for designing.**

# INTRODUCTION

**THE SOVEREIGNTY OF KNOWLEDGE**

Welcome to the era of the **Hologram Education Institute (HEI)**—a defining inflection point in the evolution of human learning, intellectual autonomy, and civilizational access to knowledge.

This moment represents more than a technological shift. It marks a **structural transformation in how civilization defines education itself.** For centuries, learning has been anchored to physical environments: schools, universities, lecture halls, and campuses constructed from material infrastructure. These systems were designed for a world characterized by geographic separation, limited communication speed, and relatively stable environmental conditions.

That world no longer exists.

## 1. The Breakdown of Traditional Educational Assumptions

Modern civilization operates under conditions that fundamentally challenge the assumptions upon which traditional education was built. Rapid population growth, accelerating urban density, climate instability, and widening economic disparity have placed increasing strain on physical educational infrastructure.

What was once scalable has become constrained. What was once accessible has become unevenly distributed.

In many regions, access to quality education is no longer determined by intellectual potential, curiosity, or ambition—but by **geographic placement and economic capacity**. This creates a structural inequality that is often invisible yet profoundly limiting: a silent barrier that restricts human development before it has the opportunity to fully emerge.

Traditional education systems, while historically transformative, are increasingly constrained by their own physicality. They require buildings to be constructed, maintained, staffed, and resourced—systems that are inherently vulnerable to disruption, whether through environmental crises, resource shortages, or infrastructural limitations.

As a result, education risks becoming **less of a universal right and more of a geographically gated privilege**.


## 2. The Emergence of a Distributed Learning Civilization

The Hologram Education Institute (HEI) emerges as a response to this structural limitation. It proposes a fundamental redefinition of what education is—and where it exists.

In the HEI paradigm, education is no longer confined to physical space. Instead, it becomes a **distributed cognitive environment**, accessible anywhere, at any time, without dependency on traditional institutional boundaries.

The classroom is no longer a room. It becomes a **volumetric experiential field**, where learners are immersed in interactive, three-dimensional knowledge environments.

The textbook is no longer static. It becomes a **living informational structure**, capable of responding to touch, movement, inquiry, and cognition.

The teacher is no longer a fixed presence bound to geography. Instead, teaching becomes a **distributed intelligence function**, accessible through a global network of knowledge systems enabled by the Subterranean Arterial Network (SAN).

In this model, learning transforms fundamentally:

- From observation —> to immersion

- From memorization —> to experience

- From standardization —> to adaptive intelligence

Education becomes not something that is delivered, but something that is **entered, navigated, and lived**.


## 3. The Core Systems That Sustain HEI

This book outlines the foundational architecture required to sustain such a system—not as theoretical speculation, but as an integrated civilizational framework.

Each subsystem plays a specific role in maintaining continuity, access, and resilience:

### Holographic Instruction Systems

Holographic Instruction restores the human dimension of education by enabling real-time presence across distance. It preserves emotional connection, dialogue, and mentorship by projecting learners and educators into shared experiential environments. This ensures that even in a fully distributed system, education remains **deeply relational rather than mechanically automated**.


### Subterranean Arterial Network (SAN)

SAN functions as the hidden infrastructural backbone of HEI. It operates as a **planetary-scale cognitive and logistical network**, transmitting data, instructional content, and system intelligence beneath the surface of the Earth.

Its subterranean design ensures resilience against surface-level instability—whether environmental, political, or infrastructural.

By embedding communication pathways deep within geologically stable layers, SAN provides continuity even under conditions of disruption.

It is not merely a data system. It is the **circulatory system of distributed civilization**.


## SHIELD Energy Infrastructure

The SHIELD Project provides the foundational energy resilience required to sustain HEI operations. Built on decentralized renewable generation systems and adaptive redistribution protocols, SHIELD ensures that educational continuity is never dependent on fragile centralized grids.

Its "Sprinkler Head" protocol dynamically reallocates energy resources during periods of stress, prioritizing critical systems such as learning environments, communication nodes, and identity authentication layers.

In this sense, SHIELD does not simply supply energy—it **stabilizes the conditions for civilization to function continuously**.


## Fiji's Cultural Legacy Framework

Embedded within HEI is a deep recognition that technological advancement alone is insufficient for sustainable civilization. Cultural memory provides identity, meaning, and ethical grounding.

Fiji's cultural legacy—along with broader Pacific knowledge systems—serves as a foundational reference point for ecological balance, communal stewardship, and environmental reciprocity.

Within HEI, cultural heritage is not preserved as static documentation. It is integrated into **living educational environments**, ensuring that identity remains active, contextual, and continuously experienced rather than archived and distant.

## 4. Toward Intellectual Sovereignty

At the center of the HEI framework lies a defining principle: **sovereignty**.

Not sovereignty in the narrow political sense, but in a deeper intellectual and existential dimension—the right of every individual to fully own their learning, their data, and their cognitive development without external restriction or institutional dependency.

In traditional systems, knowledge is often mediated through institutions that control access, validation, and certification. In the HEI model, this structure is fundamentally inverted.

Knowledge is no longer granted. It is **inherently possessed**.

Each learner becomes the sovereign owner of their educational identity, protected through decentralized systems that ensure continuity, privacy, and autonomy. This prevents the fragmentation or loss of intellectual identity within large-scale global systems.

In a world increasingly shaped by interconnected infrastructures—the "spider's web" of global technological dependency—HEI ensures that individuals remain **autonomous nodes within the system rather than passive endpoints of it**.

They are not users of knowledge.

They are its custodians.


## 5. The Foundational Shift

Ultimately, the Hologram Education Institute represents a foundational shift in civilizational design.

It redefines education as:

- A continuous environment rather than a fixed institution

- A distributed system rather than a centralized authority

- An immersive experience rather than a passive transmission

- A sovereign right rather than a managed privilege

This transformation is not incremental. It is structural.

It proposes that the future of humanity depends not merely on improving existing systems, but on **rebuilding the logic of how systems themselves are conceived**.


**Final Statement — The Beginning of a New Educational Civilization**

This is the foundation upon which the future is built:

A future where learning is no longer limited by geography. A future where knowledge is not controlled, but shared and owned. A future where education is not a privilege of place, but a **universal condition of human existence**.

And above all:

A future where intelligence is not centralized—but **collectively and sovereignly lived by every individual within it**.

# CHAPTER 1

THE PHYSICS OF HOLOGRAPHIC INSTRUCTION: HDLP
AND MIXED REALITY

The foundation of the Hologram Education Institute rests on an entirely new educational physics known as **HDLP (Holographic Dynamic Learning Projection)**. This framework does not represent an incremental improvement over existing digital learning systems, nor is it an extension of virtual or augmented reality as currently understood. Instead, it is a fundamental redefinition of how knowledge occupies space, time, and perception. Within HDLP, education is no longer treated as the transmission of information from a source to a recipient; it becomes the construction of experiential reality itself. Traditional screens function as passive surfaces that display information, while conventional virtual environments isolate the learner from physical reality. HDLP dissolves both limitations by generating fully volumetric learning fields where knowledge exists as spatially coherent, interactive, and physically coexistent environments. In this paradigm, information is no longer representation—it becomes presence, and learning is no longer observed from a distance; it is inhabited directly.

This shift in educational physics is supported by three foundational pillars that define the operational integrity of the HEI ecosystem: **data sovereignty, physical infrastructure, and energy continuity**. First, **data sovereignty** ensures that every student maintains full ownership of their intellectual development through a Decentralized Identity (DID) system. Within this structure, each learner possesses an individualized form of "Intellectual DNA," a secure and self-owned record of knowledge acquisition, cognitive growth, and cultural interaction. This guarantees that learning is not controlled or extracted by centralized institutions but remains permanently under the stewardship of the individual. Second, the system is supported by a resilient **physical infrastructure** composed of fiber-optic "Knowledge Arteries" buried deep underground. These subterranean channels are engineered for stability, cooled naturally by the Earth's thermal layers, and protected from surface-level disruptions such as environmental instability, congestion, or external interference. This design ensures uninterrupted transmission of educational and cognitive data across the global learning network. Third, the entire system is sustained by an integrated **energy architecture powered by the SHIELD PROJECT windmill grid**, reinforced by the "Sprinkler Head" safety protocol. This mechanism guarantees continuous operational uptime, adaptive energy distribution, and safeguarded system resilience, ensuring that learning environments remain stable, secure, and accessible under all conditions.

At its core, holographic instruction is not projection in the conventional optical sense, but a spatial reconstruction of knowledge into experiential reality. A lesson is no longer something delivered to the learner; it is something the learner enters, navigates, and influences through movement and interaction. The distinction between observer and environment is removed, replaced by a unified cognitive and physical learning state in which understanding emerges through direct participation. In this system, the boundary between learning and living begins to dissolve, allowing education to function as an immersive extension of reality itself. Knowledge is no longer abstracted through symbols alone; it is encountered as structure, space, and dynamic process.

Within the HEI architecture, traditional classrooms cease to exist as fixed physical containers and are replaced by adaptive projection fields. These fields are dynamically generated learning environments capable of reconstructing entire domains of knowledge at any scale, responding in real time to curriculum requirements, learner interaction, and environmental variables. Unlike static simulations, these projection fields are responsive realities that adjust their structure based on engagement, cognition, and instructional intent. A biological cell, for example, is no longer confined to diagrams or microscopic imagery; it becomes a fully navigable micro-environment where organelles function as spatial structures, and molecular processes unfold as observable and interactive phenomena. A wind turbine is no longer reduced to schematic representation; instead, it becomes a full-scale mechanical environment in which learners can move through blades, generators, and energy conversion systems, observing the transformation of force into electricity as a physical experience. Likewise, a historical village is no longer reconstructed through text or passive visualization; it becomes a fully inhabitable cultural ecosystem in which learners engage directly with architecture, language, social structures, and historical dynamics as living systems. In this model, students are no longer positioned in front of knowledge as passive recipients; they are placed inside knowledge itself, becoming active participants within the architecture of what they are learning. This shift represents a transition from observational education to immersive epistemology, where understanding is achieved through presence rather than abstraction.

To ensure that holographic instruction extends beyond visual immersion and becomes fully embodied cognition, HDLP integrates advanced haptic feedback systems and mixed reality interfaces that bind perception to physical interaction. Learning within this system is not limited to seeing or hearing; it is extended to touching, resisting, balancing, and physically engaging with reconstructed environments. Force-feedback haptic gloves simulate texture, density, resistance, temperature, and structural integrity, allowing learners to physically experience the properties of both natural and engineered systems. Motion tracking systems interpret natural body movement as direct interaction with the learning environment, eliminating the need for artificial controllers and enabling seamless physical engagement. Environmental response sensors continuously analyze learner behavior, cognitive load, and interaction patterns, dynamically adjusting simulation complexity to maintain optimal learning conditions. Together, these systems create a closed-loop cognitive-physical learning cycle in which perception drives action, and action reinforces cognition. This loop is essential, as it transforms learning from a purely mental activity into a neurological-physical integration process, strengthening memory retention through embodied experience. Knowledge is no longer stored as abstract information alone; it is encoded through movement, sensation, and spatial memory embedded within lived experience.

The HDLP system is structured with scalable implementation tiers to ensure accessibility across varying levels of infrastructure, institutional capacity, and learner needs. The first tier, known as the Desktop Holobox, provides individual immersion units that function as personal-scale learning environments. These compact systems allow learners to engage in individualized curricula, explore private spatial simulations, and interact securely with SAN-connected educational content at their own pace. The Holobox serves as the foundational gateway into volumetric learning, making immersive education accessible beyond institutional boundaries and into personal learning spaces. The second tier expands HDLP into room-scale projection systems that transform entire physical spaces into shared immersive classrooms. Within these environments, multiple learners can participate simultaneously in synchronized simulations, guided by instructors who can manipulate and scale the learning field in real time. These systems support collaborative problem-solving, collective exploration, and shared experiential learning, restoring the social dimension of education while preserving individualized interaction within the same spatial framework. The third tier, the most advanced implementation, is the AR Overlay System, which integrates learning directly into the physical world through lightweight augmented reality interfaces such as wearable glasses or advanced retinal projection systems. This allows instructional layers to be embedded seamlessly into real environments, enabling learners to interact with digital information superimposed on physical objects, receive real-time contextual guidance during practical tasks, and transition fluidly between physical and simulated realities without disruption. In this configuration, education is no longer confined to designated spaces but becomes continuous, adaptive, and fully integrated into everyday life.

Together, these three implementation tiers establish a unified educational framework in which learning is no longer dependent on geography, institutional access, or physical infrastructure. Within the HDLP paradigm, education becomes spatial rather than symbolic, experiential rather than observational, adaptive rather than static, and integrated rather than isolated. This represents the emergence of a new educational physics in which knowledge is no longer transmitted across distance but constructed as a shared and inhabitable reality. Within the Hologram Education Institute, learning is no longer something that happens to the student from the outside. It becomes something the student enters, inhabits, and ultimately transforms through direct participation within the fabric of constructed experience.

# CHAPTER 2

INTEGRATING THE SAN: THE SUBTERRANEAN NERVOUS

The Subterranean Arterial Network (SAN) constitutes the hidden cognitive backbone of the Hologram Education Institute, operating as far more than a conventional data transmission infrastructure. Within the HEI framework, SAN is conceptualized as the subterranean nervous system of global knowledge—an integrated, living lattice through which information, identity, and learning experiences continuously circulate. It replaces the traditional paradigm of centralized servers and institution-bound databases with a distributed geological intelligence system, where knowledge is no longer stored in isolated locations but dynamically flows through interconnected subterranean pathways. These deep-earth fiber-optic conduits, referred to as "Knowledge Arteries," form the primary medium of transmission, linking learners, HDLP environments, and institutional systems across vast and diverse geographic regions.

These underground arteries are engineered with environmental symbiosis at their core. By utilizing the Earth's natural thermal stability, SAN achieves passive cooling and physical shielding without reliance on external climate control systems. The subterranean positioning of the network protects it from surface-level volatility such as climate disruption, urban congestion, electromagnetic interference, and infrastructural overload. In this sense, the Earth itself becomes an active participant in the preservation of knowledge, functioning as both a stabilizing force and a protective medium. Data does not merely travel through the Earth—it is sustained by it.

At the foundation of SAN lies the principle of **Data Sovereignty**, a structural and ethical guarantee that ensures every learner retains absolute ownership of their "Intellectual DNA." Through Decentralized Identity (DID) frameworks, each individual exists within the system as a self-governing data entity, where educational records, cognitive development patterns, simulation histories, and learning trajectories are cryptographically secured and distributed across SAN nodes. This design ensures that identity is not stored in a singular authority or centralized institution but is instead fragmented, encrypted, and sovereign by default. As a result, the system cannot extract, alter, or appropriate learner identity; it can only recognize and interact with it under conditions defined by the individual. Within SAN, participation is never equivalent to surrendering control—instead, it is an act of reinforced autonomy within a protected cognitive ecosystem.

Operationally, SAN functions as a distributed intelligence grid composed of regional knowledge nodes, layered redundancy systems, and adaptive routing architectures. Each node operates both independently and collectively, enabling the system to maintain uninterrupted function even under extreme environmental or systemic stress. Redundant failover pathways ensure that no single disruption can compromise the integrity of the network, while adaptive routing algorithms continuously evaluate optimal transmission paths in real time. When obstacles arise—whether physical, digital, or geopolitical—the system does not fail; it reroutes. In doing so, SAN eliminates the traditional concept of a single point of failure and replaces it with a self-healing, self-balancing infrastructure that evolves in response to changing conditions. This guarantees that education delivery, identity verification, and cognitive interaction remain continuous across all participating regions.

A defining strength of SAN lies in its deep integration with the **SHIELD energy architecture**, which serves as the sustaining force behind all network operations. The SHIELD PROJECT is a decentralized energy ecosystem composed primarily of renewable windmill clusters, subterranean energy conduits, and distributed stabilization units that mirror the geographic logic of SAN itself. This mirrored structure creates a dual-layer system in which data and energy are structurally aligned, ensuring that neither can exist in a state of instability without affecting the other. Through this alignment, SAN becomes the informational circulatory system, while SHIELD functions as the sustaining metabolic force that keeps it alive and operational.

Within SHIELD, the "Sprinkler Head" safety protocol plays a critical regulatory role. This mechanism dynamically redistributes energy in response to fluctuations in demand, environmental strain, or system overload. Rather than allowing localized failure or depletion, the system diffuses energy across multiple nodes, prioritizing essential operations such as HDLP immersive learning environments, SAN identity authentication layers, and real-time communication channels. This ensures uninterrupted operation even during peak load conditions or external disruption. In effect, SHIELD guarantees not only continuous functionality but also systemic safety, ensuring that educational environments remain stable, responsive, and secure at all times.

The integration of SAN and SHIELD produces a unified infrastructural organism in which data, energy, and identity operate as interdependent components of a single systemic architecture. SAN carries the flow of intelligence, SHIELD sustains its vitality, and together they establish the invisible foundation upon which the HEI ecosystem is constructed. Within this framework, education is no longer constrained to physical institutions or centralized systems; it becomes a continuous, globally distributed process embedded within the Earth itself. The planet is no longer external to the system—it is structurally embedded within it, acting as both medium and guardian of civilizational knowledge.

Ultimately, SAN represents more than a technological network. It functions as the circulatory system of a sovereign learning civilization— one in which knowledge is perpetually in motion, identity remains fundamentally protected, and education persists without interruption. Through its integration with SHIELD and its foundation in Data Sovereignty, SAN establishes a new infrastructural

philosophy: one where learning is not delivered from center to periphery, but continuously sustained through a living, distributed system embedded within the Earth itself.

# CHAPTER 3

ENERGY SECURITY: THE SHIELD PROJECT AND THE SPRINKLER PROTOCOL

The evolution of the Hologram Education Institute (HEI) culminates in the emergence of **Latency-Free Cognition**, a condition in which the traditional boundaries between perception, computation, and understanding are effectively dissolved. In conventional systems, learning is constrained by sequential processing—information is requested, transmitted, interpreted, and then finally understood. Within HEI, however, this delay is eliminated through the convergence of SAN's subterranean intelligence infrastructure, SHIELD's distributed energy architecture, and a fully sovereign identity layer governed by Data Sovereignty protocols. In this environment, education ceases to be a delayed transmission of knowledge and becomes an immediate, continuous experience of cognitive immersion.

At its deepest level, Latency-Free Cognition is not merely a technological achievement but a restructuring of experiential reality. The learner is no longer positioned outside the system, receiving knowledge as an external input. Instead, the learner exists as an active participant within a living cognitive field where information responds instantly to intention, curiosity, and interaction. Every question becomes an activation signal within SAN, and every response is rendered not as a static answer but as a dynamic experiential environment. Knowledge is no longer observed—it is inhabited.

This transformation is made possible by the seamless integration of SAN's subterranean "Knowledge Arteries," which function as ultra-low latency conduits between cognitive intent and system response. These pathways are supported by distributed edge intelligence nodes embedded throughout the network, allowing computation to occur in proximity to the learner's interaction point. This architectural proximity effectively collapses physical and computational distance, reducing the perceptual gap between thought and manifestation. As a result, holographic environments respond not in milliseconds of delay, but in what is experienced as immediate continuity—where intention and system output appear indistinguishable.

However, Latency-Free Cognition is not defined solely by speed. It is defined by **continuity of consciousness within the learning process**. In traditional digital environments, interruptions in processing create fragmentation of attention and understanding. In the HEI model, SAN eliminates this fragmentation by ensuring uninterrupted cognitive flow across all learning environments. When a learner transitions between concepts, simulations, or disciplines, SAN preserves contextual continuity, allowing knowledge structures to remain active, responsive, and interconnected. Learning becomes non-linear, multidimensional, and simultaneously accessible across multiple conceptual layers.

The SHIELD energy architecture plays an equally essential role in maintaining this uninterrupted cognitive continuity. Because Latency-Free Cognition depends on constant system responsiveness, SHIELD ensures that no energy fluctuation disrupts computational or sensory output. Its decentralized renewable grid, composed of wind-based energy clusters and subterranean distribution channels, mirrors the spatial logic of SAN. This mirroring creates a synchronized infrastructure where energy flow and data flow are structurally aligned. Through this alignment, SHIELD functions as the metabolic engine of the system, continuously sustaining the cognitive field in which learning occurs.

The "Sprinkler Head" safety protocol further enhances this stability by dynamically managing energy distribution across the network. During periods of heightened cognitive demand—such as complex simulations, multi-layered HDLP environments, or large-scale synchronized learning sessions—SHIELD automatically redistributes energy to prioritize critical systems. This ensures that cognitive performance remains stable even under extreme system load, preserving the integrity of real-time learning without degradation or interruption.

Within this architecture, cognition itself is redefined as a distributed phenomenon rather than a localized neurological event. The learner's mind is no longer the sole processor of information but a node within a broader intelligence ecosystem. SAN extends cognitive reach beyond biological limitations, allowing perception to interface directly with structured data environments. In this state, thinking becomes spatial, interactive, and environmental. Concepts are no longer abstract constructs held internally; they become navigable spaces within a holographic learning field.

This shift introduces a new educational dynamic in which **experience replaces instruction**. Instead of being taught about systems, learners engage directly with functional simulations of those systems. A mathematical principle is not explained—it is experienced as a manipulable structure. A historical event is not described—it is rendered as a reconstructed experiential timeline. Scientific principles are not memorized—they are interacted with as living models governed by real-time feedback. Through this method, understanding emerges organically through participation rather than passive reception.

Despite its immersive depth, Latency-Free Cognition remains fully governed by the principles of **Data Sovereignty**. Every cognitive interaction within the system is encrypted, authenticated, and attributed exclusively to the individual learner. No data is extracted without permission, and no experience is dissociated from its rightful owner. This ensures that even in a fully integrated cognitive environment, autonomy is preserved as a foundational principle. The system expands perception without compromising ownership, and enhances intelligence without diminishing identity.

As this system matures, the distinction between learning and living begins to dissolve. The HEI ecosystem becomes a continuous cognitive environment in which education is no longer scheduled or compartmentalized but constantly active. Learners exist within an ongoing stream of structured experience where knowledge is perpetually accessible and immediately responsive. In this sense, Latency-Free Cognition represents not just an advancement in educational technology, but a redefinition of what it means to perceive, to understand, and to exist within a knowledge-based civilization.

Ultimately, within the unified framework of SAN, SHIELD, and Data Sovereignty, Latency-Free Cognition establishes a new paradigm: a world in which education is no longer delayed by systems, constrained by infrastructure, or limited by geography. Instead, it becomes an immediate, sovereign, and continuous state of awareness—where learning and consciousness are inseparable, and where understanding is always already present.

# CHAPTER 4

EARLY CHILDHOOD: THE FIJI HISTORY IMMERSIVES

The Early Childhood framework within the Hologram Education Institute (HEI) is fundamentally redefined through the **Fiji History Immersives**, a foundational learning architecture designed to anchor early cognitive development in cultural memory, identity formation, and experiential heritage awareness. Unlike conventional early education models, which often rely on abstract symbols, simplified narratives, and fragmented historical references, this system introduces learning as **direct lived experience from the earliest stages of consciousness formation**. Within this paradigm, education is no longer the passive absorption of historical facts but the active participation within reconstructed cultural realities. The Fiji History Immersives function simultaneously as a pedagogical environment and a cultural preservation infrastructure, ensuring that Fiji's layered histories are not merely archived but continuously reactivated as living, interactive systems accessible to each new generation.

At the core of this framework is the transformation of historical knowledge into **age-adaptive immersive cognitive environments** powered by HDLP holographic instruction and SAN-supported reconstruction architecture. Through the Subterranean Arterial Network (SAN), fragmented historical data, oral traditions, ecological records, and cultural artifacts are synthesized into coherent experiential simulations that replicate not only physical environments but also social, emotional, and cultural dynamics. Children are placed within these environments as active participants rather than external observers, allowing them to experience history as a living continuum rather than a static timeline. Villages, coastal settlements, ceremonial grounds, and ecological systems are reconstructed with high contextual fidelity, enabling learners to interact with architecture, community roles, environmental cycles, and cultural practices in real time.

Within these immersive systems, learning becomes **multisensory, embodied, and relational**. A child introduced to traditional Fijian navigation, for example, does not simply study diagrams or read descriptions of ocean voyaging. Instead, they experience the ocean as a dynamic field, interpret star patterns through guided interaction, and engage with decision-making scenarios modeled after ancestral navigators. The wind, current, and celestial movements are not presented as abstract concepts but as interactive environmental forces that respond to learner actions. This transforms cognition from symbolic understanding into **spatial and experiential encoding**, where knowledge is deeply integrated into memory through sensory engagement, emotional association, and environmental interaction.

The Fiji History Immersives also function as a **living cultural transmission system**, enabling intergenerational knowledge transfer with unprecedented fidelity and depth. Elders, cultural custodians, historians, and community knowledge holders contribute directly to the SAN knowledge ecosystem, where their oral histories, lived experiences, and ancestral teachings are encoded as foundational data layers within immersive simulations. These contributions are not treated as secondary interpretations but as primary epistemological sources that actively shape the structure, behavior, and authenticity of the learning environments. In this way, history is not externally authored or abstractly curated; it is continuously co-created by those who embody it, ensuring that cultural integrity is preserved while remaining dynamically adaptable within the educational system.

Within the HEI framework, this process establishes a new form of **cultural sovereignty through knowledge architecture**. By embedding indigenous knowledge systems directly into SAN's distributed intelligence network, cultural heritage is protected from distortion, dilution, or external reinterpretation. Instead, it is preserved as a living system that evolves organically through participation while maintaining continuity with its origins. Each interaction within the immersive environment reinforces cultural identity, ensuring that learners do not inherit history as detached information but as an active, lived inheritance embedded within their cognitive development.

Beyond historical immersion, the Fiji History framework integrates deeply with **ecological cognition and environmental systems education**, reflecting the inseparable relationship between Fijian culture and its natural environment. Children are introduced to ecological interdependence through fully interactive simulations of coral reef systems, rainforest ecosystems, coastal resource networks, and climate cycles. These environments are not simplified educational models but functionally accurate ecological systems that respond dynamically to learner interaction. As children explore these environments, they observe the direct consequences of environmental balance, disruption, and restoration, developing an intuitive understanding of ecological responsibility from an early developmental stage.

This ecological integration ensures that cultural education and environmental awareness are not treated as separate domains but as a unified cognitive framework. Learners come to understand that cultural survival is directly linked to environmental stewardship, and that the continuity of heritage depends on the sustainability of natural systems. Through repeated experiential engagement, stewardship becomes not an abstract moral concept but a lived cognitive principle embedded within perception itself.

The design of the Fiji History Immersives is carefully calibrated for **early cognitive and emotional development**, ensuring that complexity is introduced in a structured and adaptive manner. The system continuously responds to learner behavior, emotional engagement, attention patterns, and cognitive readiness, adjusting the depth and intensity of experiences in real time. In early developmental stages, learning is primarily sensory, narrative-driven, and exploratory, emphasizing recognition, curiosity, and emotional connection. As cognitive capacity matures, the system gradually introduces layered historical causality, social systems analysis, environmental interdependence, and cultural dynamics, allowing learners to develop progressively deeper forms of understanding without cognitive overload.

This adaptive architecture ensures that education evolves in alignment with the learner's developmental trajectory. Rather than enforcing uniform progression, the system supports individualized cognitive pathways, allowing each child to engage with history and culture at a pace that reflects their unique cognitive and emotional development. This creates a personalized continuum of learning in which understanding is continuously reinforced, expanded, and contextualized through immersive experience.

Ultimately, the Fiji History Immersives establish early childhood education as a **foundation of cultural continuity and sovereign identity formation within the HEI ecosystem**. By embedding historical memory, ecological awareness, and cultural participation into immersive learning environments, the system ensures that children do not grow disconnected from their heritage but instead develop a deeply embodied relationship with it. Education becomes not only a mechanism for intellectual development but a process of identity formation rooted in lived continuity.

Within this framework, the past is no longer a distant record to be studied but a living system to be experienced, preserved, and carried forward. The Fiji History Immersives transform early education into a continuously active cultural archive—one in which history is not remembered from afar, but re-lived, re-integrated, and re-embodied within the evolving consciousness of each new generation.

# CHAPTER 5

SKILLED TRADES: HAPTIC OVERLAY AND VIRTUAL APPRENTICESHIPS

The Skilled Trades Framework within the Hologram Education Institute (HEI) represents the operational backbone of applied learning, where abstract knowledge is transformed into **embodied competence through immersive, simulation-driven apprenticeship systems**. In traditional educational models, vocational training is often separated from academic theory, requiring physical infrastructure, limited access to tools, and extended real-world apprenticeships that depend heavily on geographic location and industry availability. This separation creates inefficiencies, unequal access, and prolonged skill acquisition timelines. Within the HEI ecosystem, these limitations are fundamentally eliminated. The integration of Haptic Overlay Systems, SAN-supported distributed intelligence, and HDLP immersive environments enables learners to acquire technical mastery in controlled, adaptive, and fully responsive learning fields where practice, repetition, and refinement occur at accelerated cognitive and physical scales.

At the core of this framework is the **Haptic Overlay Architecture**, a system designed to convert holographic instructional environments into fully interactive sensory fields. Through advanced force-feedback interfaces, motion capture arrays, and multi-layered sensory augmentation systems, learners are able to physically engage with virtual tools, machinery, and industrial systems as though they exist in tangible reality. The environment does not simulate interaction—it **reproduces**

**the physical laws of interaction itself**, allowing users to experience weight, resistance, vibration, torque, heat feedback, and material response in real time. A learner training in mechanical engineering, for instance, can assemble a turbine engine within a holographic industrial bay, feeling the tightening resistance of bolts, the alignment tension of components, and the kinetic feedback of rotational systems. Similarly, learners in electrical engineering can interact with full-scale simulated power grids, experiencing current flow, load balancing, circuit failure cascades, and system recovery processes without exposure to physical danger. This ensures that skill development is not merely cognitive but deeply **neurological and procedural**, embedding technical intuition directly into motor memory and decision-making pathways.

The effectiveness of this system lies in its ability to synchronize **sensory experience with cognitive encoding**. Unlike traditional learning environments where theory and practice are separated by time and access, the HEI model fuses instruction and execution into a single continuous loop. Every action taken within the haptic environment produces immediate feedback, reinforcing learning through real-time correction and adaptation. This creates a closed cognitive circuit where understanding is continuously refined through doing, and doing is continuously informed by understanding. Over time, this results in accelerated mastery, as learners develop instinctive familiarity with complex systems long before encountering them in physical reality.

Extending this framework are **Virtual Apprenticeships**, which replicate entire industrial ecosystems within SAN-connected HDLP environments. These apprenticeships simulate full-scale operational domains such as construction sites, manufacturing plants, logistics networks, energy infrastructure systems, and advanced technical workshops. Within these environments, learners assume structured occupational roles that mirror real-world professional hierarchies and responsibilities. They may operate as junior technicians, system analysts, field engineers, or project coordinators depending on their developmental stage and training pathway. Each role is embedded within a dynamic operational simulation that requires collaboration, decision-making, and system awareness.

Unlike conventional apprenticeships, these environments are not constrained by material scarcity, physical safety risks, or operational downtime. Instead, they are **fully generative systems capable of producing infinite scenario variations**, including rare, high-pressure, or failure-critical conditions that would be difficult or impossible to recreate in physical industries. A learner in a construction simulation, for example, may be exposed to structural stress failures, weather-induced system instability, or material shortages, requiring rapid adaptive problem-solving. In energy system training environments, learners may manage grid overloads, cascading failures, or emergency rerouting scenarios that test both technical knowledge and cognitive resilience. Through this approach, training moves beyond routine competence into **adaptive expertise under complexity and uncertainty**.

The integration of SAN's distributed intelligence layer with HDLP immersive environments enables these simulations to function as **collective operational ecosystems**. Multiple learners can participate simultaneously within shared industrial environments, coordinating tasks, managing system outputs, and responding to evolving conditions as interconnected teams. These collaborative simulations mirror real-world industrial interdependence, where no system operates in isolation. Each participant's actions influence the broader environment, creating a feedback-rich ecosystem where communication, coordination, and systems thinking become essential competencies. SAN continuously synchronizes these interactions in real time, ensuring coherence across all participants regardless of scale or complexity.

Performance within these environments is continuously analyzed through embedded cognitive and operational feedback systems. The HEI framework evaluates not only task completion but also decision pathways, adaptability under pressure, efficiency of execution, and collaborative effectiveness. Based on this analysis, simulations dynamically adjust in real time—introducing new constraints, increasing complexity, or modifying environmental conditions to challenge the learner's evolving skill level. This creates a **self-scaling educational system**, where difficulty is not static but continuously aligned with individual and collective performance trajectories. Learning becomes an adaptive process rather than a fixed curriculum, ensuring constant developmental progression.

A critical transformation introduced by this framework is the **redefinition of safety in vocational education**. In traditional systems, exposure to high-risk environments—such as electrical grids, heavy machinery, chemical processes, or structural engineering systems—is inherently limited due to physical danger. Within HEI, all such risks are fully simulated within controlled digital environments governed by SHIELD-stabilized infrastructure. This allows learners to engage directly with hazardous scenarios without physical consequence, enabling repeated experimentation, failure analysis, and iterative correction. Failure is no longer a terminal outcome but a **data-rich learning event**, providing immediate insight into system behavior and decision impact.

This elimination of physical risk fundamentally alters the psychological relationship between learners and complexity. Without fear of permanent consequence, learners are more willing to explore unconventional solutions, test hypotheses, and engage in high-level problem-solving behaviors. Over time, this cultivates **resilience, creativity, and systems confidence**, qualities essential for mastery in real-world technical environments. The result is a generation of practitioners who are not only technically proficient but cognitively flexible and adaptive under pressure.

Ultimately, the Skilled Trades Framework represents the point at which education transitions into **direct operational capability**. Through the fusion of haptic interaction, SAN-driven simulation intelligence, and SHIELD-supported system stability, vocational learning becomes indistinguishable from real-world execution. Students are no longer separated from industry—they are embedded within its simulated equivalent from the earliest stages of training. In this model, education is not preparation for work; it is **continuous participation in engineered systems of production, infrastructure, and innovation**.

Within the HEI ecosystem, Chapter 5 establishes skilled trades as a foundational pillar of civilizational development. It elevates vocational education from a secondary educational track into a core mechanism of technological continuity, ensuring that every learner emerges not only with knowledge, but with the **operational fluency required to sustain,**

**build, and evolve complex systems in the real world and beyond it.**

# CHAPTER 6

THE PHD LEVEL: MANAGING GLOBAL RESOURCE EQUILIBRIUM

The PhD-level framework within the Hologram Education Institute (HEI) represents the highest tier of cognitive, ethical, and operational development within the entire educational ecosystem. At this stage, education transcends disciplinary mastery and transitions into the domain of **planetary stewardship**, where learners are no longer engaged in studying systems as external observers, but are actively operating within them as responsible custodians of global complexity. This level marks the point at which learning becomes indistinguishable from governance simulation, and knowledge evolves into a direct instrument for maintaining the stability, continuity, and resilience of civilization itself.

The defining principle of this framework is *Global Resource Equilibrium*, a dynamic model in which human intelligence is applied not to isolated optimization problems, but to the continuous harmonization of interconnected planetary systems. These systems include energy distribution networks, ecological balances, food production chains, water resource management, population dynamics, technological infrastructure, and economic flow systems. Within this paradigm, no system exists independently; every adjustment in one domain produces cascading effects across all others. The role of the learner, therefore, is not to maximize efficiency in a single sector, but to maintain equilibrium across an entire living planetary architecture that is constantly evolving.

At the core of this framework is the integration of **SAN (Subterranean Arterial Network)** and **HDLP (Holographic Dynamic Learning Projection)** into a unified planetary simulation environment. Within this system, PhD-level learners are immersed in continuously updated, data-synchronized replicas of global infrastructure and environmental systems. These simulations are fed in real time by distributed sensor networks, ecological monitoring stations, energy grid telemetry, agricultural output systems, and socio-economic data streams. The result is not a static academic model but a **living digital twin of planetary behavior**, continuously evolving in parallel with real-world conditions. Learners engage with this environment not as observers, but as operational agents capable of influencing system behavior through structured interventions.

Within these simulations, learners are assigned governance-level roles that require them to interpret complex interdependencies and execute decisions with far-reaching consequences. A single adjustment—such as modifying energy distribution in a specific region—can trigger cascading effects across agricultural productivity, urban infrastructure stability, transportation logistics, environmental stress levels, and economic performance indicators. This demands a fundamental shift in cognition: from linear problem-solving to **multi-layered systems reasoning**, where every decision must account for immediate, intermediate, and long-range systemic impacts simultaneously.

The concept of equilibrium within HEI is deliberately defined as **non-static and continuously adaptive**. Unlike traditional models that assume stability as a fixed endpoint, this framework recognizes that global systems exist in perpetual flux, shaped by environmental variability, technological evolution, consumption behavior, and socio-political dynamics. As a result, there is no final optimized state. Instead, learners are trained to maintain **dynamic balance within instability**, ensuring that no subsystem collapses under the pressure of another. This requires a synthesis of analytical intelligence, predictive modeling, ethical reasoning, and adaptive decision-making, all operating simultaneously within a unified cognitive framework.

To support this complexity, HEI incorporates advanced **predictive intelligence layers** that function as temporal simulation engines. These systems generate multi-horizon projections that allow learners to observe the potential consequences of decisions across short-term, mid-term, and long-term timelines. For example, a decision affecting water allocation in one geographic region may reveal downstream effects on agricultural yields, population migration patterns, economic stability, and geopolitical tension over extended simulated periods. These projections are not deterministic forecasts but probabilistic scenario landscapes, designed to train learners in **anticipatory governance**, where decisions are evaluated not only by immediate outcomes but by their systemic evolution over time.

Within this structure, learners are constantly exposed to the complexity of **causal interdependence**. Every intervention becomes part of a feedback loop that influences multiple domains simultaneously. This cultivates a deep cognitive understanding that no system is isolated, and that every action contributes to a broader network of global consequences. Over time, learners develop the ability to perceive systems holistically, recognizing patterns of vulnerability, resilience, and interconnection that are invisible within traditional educational frameworks.

Collaboration is a foundational requirement at this level. PhD-level learners operate within distributed intelligence teams connected through SAN's global coordination infrastructure. Each participant is assigned responsibility for a specific domain—such as energy systems, ecological management, infrastructure stability, or economic modeling—yet all domains remain interdependent within the shared simulation environment. This structure creates a form of **distributed governance learning**, where decision-making authority is decentralized and emergent rather than hierarchical. Leadership is not assigned; it is earned through demonstrated capacity to understand system-wide equilibrium and coordinate effectively across domains.

This collaborative structure fosters a new form of intellectual interaction that can be described as **systems diplomacy**. Learners must continuously negotiate trade-offs, reconcile competing priorities, and align objectives across multiple domains that may be in tension with one another. For instance, increasing industrial output may improve economic indicators but place strain on energy systems and ecological balance. Resolving such tensions requires not only technical expertise but also communication, ethical reasoning, and strategic foresight. In this way, governance becomes a shared cognitive process rather than a centralized authority structure.

Ethics is not treated as an external constraint but as a **core operational parameter embedded within the simulation itself**. Every decision carries explicit moral, ecological, and social consequences, which are rendered visible through system feedback. Learners are confronted in real time with the implications of inequality, resource distribution imbalance, environmental degradation, and infrastructural prioritization. This ensures that technical optimization cannot be separated from ethical responsibility. The system is designed to continuously surface these tensions, requiring learners to develop an integrated moral-technical intelligence capable of balancing efficiency with equity, innovation with sustainability, and progress with preservation.

As learners advance through this framework, their cognitive role shifts from problem-solving to **system stewardship**. They are no longer reacting to challenges but actively shaping the conditions under which challenges emerge. This represents a transition from operational intelligence to structural intelligence—the ability to influence not only outcomes but the architecture of the systems that generate those outcomes. In this sense, education at the PhD level becomes indistinguishable from participation in the ongoing maintenance of civilization itself.

Ultimately, the PhD-level framework of HEI redefines the meaning of advanced education. It is no longer the endpoint of academic specialization, but the entry point into planetary responsibility. Learners at this stage are not merely experts within their fields; they are **custodians of interconnected global systems**, trained to maintain equilibrium across the fragile architecture of modern civilization. Within this paradigm, knowledge becomes inseparable from governance, and learning becomes an ongoing act of stewardship.

In managing global resource equilibrium through SAN, SHIELD-supported infrastructure, and HDLP simulation environments, learners are not preparing for a theoretical future. They are actively participating in the continuous construction, stabilization, and evolution of that future in real time—within a living, adaptive model of the world itself.

# CHAPTER 7

## DIGITAL SOVEREIGNTY: THE BLOCKCHAIN LEDGER OF KNOWLEDGE

The Digital Sovereignty framework within the Hologram Education Institute (HEI) establishes the fundamental principle that all knowledge, identity, and educational progression must remain permanently owned by the learner who generates them. In conventional educational systems, intellectual history is typically mediated through centralized institutions—universities, certification bodies, and governmental archives—that act as custodians of academic identity. Within such structures, access to credentials, records, and intellectual outputs can be restricted, delayed, altered, or even lost due to institutional policy, geopolitical instability, or administrative fragmentation. The HEI model removes this dependency entirely by embedding sovereignty directly into the architecture of learning itself through a **Blockchain-based Ledger of Knowledge**, where ownership is not granted by institutions but inherently generated through participation.

At the core of this system is the principle of **distributed identity**, a cryptographically secured identity architecture that ensures each learner exists as a persistent, verifiable entity across the entire HEI ecosystem. This identity is not stored in a single centralized repository but is instead distributed across SAN (Subterranean Arterial Network) nodes and continuously synchronized through blockchain validation protocols. Every interaction within the system—whether it occurs in early childhood immersion environments, skilled trade simulations, or PhD-level global governance models—is recorded as a validated cognitive event. These events form a continuous, chronological chain of intellectual development, ensuring that the learner's educational history is both permanent and globally verifiable. In this structure, identity is no longer institution-dependent; it is system-native, existing independently of any single authority or geographic jurisdiction.

The **Blockchain Ledger of Knowledge** functions as the structural backbone of intellectual permanence within HEI. Each learning event is encoded as a cryptographically sealed "knowledge block," containing not only the outcome of a learning interaction but also contextual metadata such as decision pathways, simulation environments, problem-solving strategies, and collaborative contributions. These blocks are linked sequentially, forming an immutable cognitive timeline that represents the full scope of a learner's intellectual evolution. Once recorded, these entries cannot be altered retroactively without distributed consensus across the network, ensuring the integrity and authenticity of all educational records. This creates a system in which knowledge is not merely acquired and stored, but permanently attributed and structurally preserved as part of a lifelong cognitive architecture.

Beyond record-keeping, this ledger introduces a new dimension of **intellectual provenance and authorship integrity**. Within HEI's highly collaborative and simulation-rich environments, learners frequently generate solutions, designs, cultural reconstructions, and system interventions that contribute to shared knowledge ecosystems. The blockchain layer ensures that every intellectual contribution is cryptographically timestamped and attributed to its originator at the moment of creation. Whether a learner engineers a solution in a SHIELD energy simulation, designs infrastructure within SAN-connected environments, or reconstructs historical narratives in HDLP cultural systems, their contribution is permanently recorded as verifiable intellectual property. This creates a transparent and traceable ecosystem of innovation in which creativity is both protected and recognized without requiring centralized oversight.

Importantly, this system is designed not to centralize authority over knowledge validation, but to **distribute trust across the entire network**. Instead of relying on a single governing institution to verify educational legitimacy, HEI uses decentralized consensus mechanisms that operate across SAN's subterranean infrastructure and blockchain validation layers. This ensures that no single entity can alter, erase, or manipulate educational records. Validation emerges collectively from the system itself, creating a self-regulating epistemic environment where truth, authorship, and achievement are continuously verified through distributed agreement rather than institutional control.

A defining feature of Digital Sovereignty within HEI is the **decoupling of identity from institutional dependency**. In traditional systems, a learner's academic identity is often tied to enrollment within specific institutions, meaning that certification, recognition, and academic continuity are contingent upon institutional authority. Within HEI, identity exists independently of any institution and is instead anchored directly in the Blockchain Ledger of Knowledge. This allows learners to carry their entire cognitive and educational history across all environments within the ecosystem, whether they are engaging in SAN-connected simulations, HDLP immersive environments, or real-world augmented learning interfaces. Their achievements are universally portable, continuously verifiable, and permanently accessible across the entire system architecture.

This portability introduces a new form of **educational continuity across space, time, and systemic change**. Learners are no longer vulnerable to disruptions caused by relocation, institutional closure, policy changes, or technological transitions. Their intellectual identity persists as a continuous, system-independent construct that remains intact regardless of external conditions. This ensures that education becomes a lifelong, uninterrupted trajectory rather than a fragmented sequence of institutional experiences.

Security within this framework is maintained through multi-layered cryptographic systems designed to ensure integrity, resilience, and resistance to tampering. Each educational event is encrypted and mathematically linked to preceding entries, forming a continuous chain of cognitive validation that is both transparent and secure. This structure ensures that while authorized systems can access and verify educational data, no unauthorized modification or duplication is possible. The integration of SAN further strengthens this architecture by distributing ledger data across subterranean nodes, eliminating centralized vulnerabilities and significantly reducing the risk of systemic compromise. In this configuration, security is not a separate layer added to the system—it is embedded within the structural logic of knowledge itself.

Beyond its technical architecture, Digital Sovereignty introduces a profound **philosophical redefinition of education and identity**. Learning is no longer viewed as a temporary phase that culminates in certification or graduation; it becomes a continuous extension of personal existence. Every simulation completed, every problem solved, every system designed, and every insight generated becomes part of an evolving cognitive signature that defines the learner's presence within the HEI ecosystem. Education is therefore transformed from a sequence of achievements into a permanent record of lived intellectual experience—an evolving narrative of thought, action, and understanding.

Within this paradigm, sovereignty takes on its most complete meaning. The learner is not simply a participant in an educational system but the **permanent author and custodian of their intellectual evolution**. Their knowledge cannot be removed from them, separated from them, or redefined without systemic consensus. Identity is no longer granted by institutions but emerges intrinsically through participation and is preserved permanently through cryptographic verification.

Ultimately, Digital Sovereignty within HEI establishes a foundational constitutional layer for the entire educational ecosystem. It ensures that knowledge cannot be detached from its creator, that identity cannot be controlled by external institutions, and that educational history cannot be erased, rewritten, or centralized. It creates a system in which learning is simultaneously secure and autonomous, structured yet self-owned, distributed yet unified in integrity.

In this framework, the Blockchain Ledger of Knowledge transcends its technical function and becomes something far greater: a **permanent architecture of intellectual freedom**. It affirms every learner as a sovereign entity within a distributed civilization of knowledge—recognized not only as a recipient of education, but as the enduring author of their own cognitive legacy, preserved across time within the interconnected architecture of SAN, SHIELD, and the global HEI learning continuum.

# CHAPTER 8

THE DATA SPRINKLER: DEFENDING INTELLECTUAL PROPERTY

The Data Sprinkler system within the Hologram Education Institute (HEI) represents a next-generation architecture for the protection, distribution, and controlled activation of intellectual property across a fully immersive educational civilization. In environments where knowledge is continuously generated through SAN-connected simulations, HDLP experiential learning layers, and SHIELD-supported infrastructural modeling, traditional intellectual property systems—built on static ownership, centralized repositories, and perimeter-based security—become fundamentally insufficient. The Data Sprinkler replaces this outdated model with a dynamic, biologically inspired system of **adaptive dispersion**, where protection is achieved not through confinement, but through intelligent fragmentation, contextual reconstruction, and controlled accessibility.

At its foundation, the Data Sprinkler operates as a **distributed containment architecture embedded within the SAN (Subterranean Arterial Network)**. Instead of storing intellectual assets—such as advanced simulations, proprietary HDLP learning environments, or high-value research frameworks—as unified, accessible datasets, the system decomposes them into modular fragments. Each fragment is individually encrypted, structurally incomplete, and contextually dependent. On its own, a fragment carries no functional value and cannot be reverse-engineered into a usable system. Only when accessed through authenticated HEI environments, validated learner identities, and secure SAN routing protocols can these fragments be temporarily recombined into coherent operational forms. This ensures that intellectual property is never exposed in a complete state outside of authorized cognitive environments, fundamentally eliminating the risk of full-system extraction.

The defining principle behind this architecture is **adaptive dispersion**, a strategy derived from natural systems that ensure survival through distributed redundancy. Much like biological ecosystems that scatter seeds across wide environments to guarantee species continuity under unpredictable conditions, the Data Sprinkler disperses intellectual assets across a vast network of subterranean nodes, each acting as a micro-fragment of a larger informational structure. This dispersion ensures that no single point of failure, breach, or compromise can result in the reconstruction or theft of complete intellectual systems. Even in the case of localized intrusion, the fragmented architecture guarantees that the exposed data remains incoherent, unusable, and mathematically non-reconstructable without full system authorization.

This model fundamentally transforms the concept of security from a static barrier into a **living ecological process**. Instead of building walls around knowledge, HEI allows knowledge to exist in a state of controlled incompleteness across multiple secure environments. Security is achieved not by preventing movement, but by ensuring that movement never results in exposure of complete systems. In this way, intellectual property behaves less like a stored object and more like a distributed organism—resilient, adaptive, and continuously self-protecting through structural dispersion.

A central component of the Data Sprinkler system is **intelligent access modulation**, which governs how fragmented intellectual assets are temporarily reconstructed within authorized environments. Access is not binary but dynamically contextual, determined by a combination of learner identity, educational progression level, SAN system role, SHIELD clearance tier, and situational learning objectives.

This creates a multi-layered access ecosystem where the same intellectual asset may exist in different degrees of completeness depending on the user's role within the system.

For example, foundational learners may only interact with abstracted or partially reconstructed versions of a complex HDLP simulation, designed to introduce core concepts without exposing full system architecture. In contrast, advanced PhD-level researchers operating within secure SAN governance environments may access full-resolution reconstructions of the same simulation, but only under real-time monitoring, temporal access constraints, and system audit validation. This ensures that intellectual property is not only protected from unauthorized duplication but is also **scaffolded educationally**, adapting its complexity to the cognitive maturity and responsibility level of the user.

The Data Sprinkler also introduces a critical innovation in **context-bound reconstruction logic**. Intellectual fragments are not universally reassembleable; instead, they are tied to specific environmental conditions, simulation parameters, and authorized learning contexts. This means that even if fragments are intercepted or copied, they cannot be reconstructed outside of their original SAN-validated environment. Each fragment carries embedded constraints that define where, when, and under what conditions it can be activated, ensuring that intellectual property remains functionally inert outside of its intended educational ecosystem.

Within collaborative HEI environments, particularly those operating through HDLP immersive systems and SAN-connected research clusters, the Data Sprinkler plays a vital role in protecting **co-generated intellectual ecosystems**. Learners and researchers frequently engage in collective simulation-building, system modeling, and cross-disciplinary innovation. Without structured protection, these shared outputs would be vulnerable to uncontrolled replication or unauthorized redistribution. The Data Sprinkler addresses this by embedding **cryptographic lineage tracking** into every modular fragment of a system.

Each fragment contains metadata that records its origin, contributors, modification history, and permissible reconstruction pathways. This creates a living audit trail of intellectual evolution, ensuring that every contribution is both traceable and attributable. At the same time, usage constraints embedded within each fragment define how it can be recombined, modified, or extended within the system. This ensures that collaborative innovation remains open and dynamic, while still preserving the integrity and authorship of each contributor's intellectual input.

Beyond its technical infrastructure, the Data Sprinkler introduces a profound philosophical shift in how intellectual property is understood within a fully immersive knowledge civilization. In traditional systems, knowledge is treated as a static asset that must be enclosed, protected, and restricted to prevent unauthorized access. Within HEI, knowledge is reconceptualized as a **living, distributable structure that must be carefully maintained through intelligent exposure rather than isolation**.

In this model, security is not achieved through permanent restriction but through controlled accessibility. Knowledge is allowed to circulate across systems, environments, and learning contexts, but never in a form that compromises its structural integrity or authorship identity. This ensures that innovation remains fluid and collaborative while still respecting the sovereignty of creators within the HEI ecosystem. Intellectual property becomes less about possession and more about **responsible stewardship of distributed knowledge systems**.

The Data Sprinkler also reinforces the broader HEI architecture by functioning as a complementary layer to SAN, SHIELD, and the Blockchain Ledger of Knowledge. While SAN provides infrastructural distribution, SHIELD manages systemic energy and operational integrity, and the Blockchain Ledger ensures identity and authorship permanence, the Data Sprinkler governs the **controlled fragmentation and reconstitution of intellectual assets**. Together, these systems form a unified ecosystem in which knowledge is simultaneously secure, portable, traceable, and dynamically accessible.

Ultimately, the Data Sprinkler system redefines intellectual property protection as a **continuous adaptive process rather than a static defensive boundary**. It ensures that knowledge cannot be stolen in complete form, cannot be reconstructed outside authorized environments, and cannot be separated from its authorship lineage. At the same time, it guarantees that knowledge remains usable, accessible, and evolving within structured educational contexts.

Within the HEI framework, the Data Sprinkler is more than a security mechanism—it is a governing philosophy of informational resilience. It reflects a world in which knowledge is too complex, too interconnected, and too continuously evolving to be locked away. Instead, it must be dispersed, protected through structure rather than isolation, and activated only within environments capable of honoring its integrity.

In this way, the Data Sprinkler transforms intellectual property protection into a living architecture—one that mirrors the complexity of cognition itself and ensures that knowledge remains both free to move and impossible to exploit outside the ecosystem that gives it meaning.

# CHAPTER 9

## DECENTRALIZING VITI LEVU: THE NORTH-SOUTH & EAST-WEST FREEWAYS

The infrastructure model of the Hologram Education Institute (HEI) extends beyond digital ecosystems, educational architecture, and data governance frameworks into the **physical reconfiguration of geography itself**. In this paradigm, land is no longer viewed as passive terrain awaiting development, but as an active structural component of civilization's cognitive, economic, and educational systems. Chapter 9 focuses on the decentralization of Viti Levu through the strategic deployment of the **North-South and East-West Freeways**, a dual-axis infrastructural system designed to fundamentally reshape movement, opportunity distribution, and regional equilibrium across the island.

Within the HEI framework, transportation is not treated as a neutral mechanism for mobility, but as a **core intelligence layer of civilization**. Roads, corridors, and transit systems become extensions of national cognition—channels through which economic activity, human development, and knowledge dissemination continuously circulate. In this model, infrastructure functions as a physical analogue to SAN (Subterranean Arterial Network), translating the principles of distributed intelligence into geographic form. The decentralization of Viti Levu therefore represents not only an engineering intervention but a systemic reconfiguration of how civilization organizes itself across space.

At the foundation of this model is the **North-South Freeway**, conceptualized as the longitudinal spine of the island's developmental architecture. This corridor establishes a continuous structural linkage between northern and southern regions, integrating agricultural zones, population clusters, resource extraction areas, and emerging industrial corridors into a single coherent mobility axis. Rather than allowing development to concentrate in isolated coastal cities, the North-South Freeway distributes economic activity along its entire length, enabling synchronized regional growth.

This axis functions as a **redistribution engine for economic and human capital**, ensuring that rural and inland regions are no longer structurally disadvantaged in comparison to coastal hubs. By enabling high-capacity, uninterrupted movement of goods, services, labor, and information, the freeway transforms spatial isolation into integrated participation. Each node along the corridor becomes part of a continuous developmental gradient rather than an isolated economic unit. In this way, the North-South axis does not simply connect regions—it **redefines their developmental relationship to one another**, ensuring that no area remains permanently peripheral to national progress.

Complementing this longitudinal structure is the **East-West Freeway**, which introduces a horizontal layer of connectivity across ecological, cultural, and economic zones. While the North-South system establishes structural continuity along the island's vertical axis, the East-West system enables lateral exchange between inland regions and coastal economies, creating a fully interconnected lattice of movement and interaction. This horizontal integration is critical for addressing historical patterns of imbalance in which coastal zones have disproportionately controlled access to trade, infrastructure, and institutional development.

Through the East-West corridors, inland communities gain direct access to coastal trade routes, educational institutions, healthcare systems, and technological hubs without requiring centralized transit through overburdened urban cores. This reduces systemic congestion while simultaneously expanding opportunity distribution across the entire island. The result is a **bidirectional flow model**, where resources move inward from coastal zones and innovation, labor, and agricultural output move outward in continuous circulation. The island becomes a self-regulating spatial system rather than a centralized hierarchy of development.

When combined, the North-South and East-West Freeways form a **dual-axis spatial intelligence grid**, effectively transforming Viti Levu into a structured lattice of interconnected development corridors. This system resembles a dynamic "spider's web" of infrastructure, where each intersection point functions as a node of concentrated activity and systemic convergence. These intersections are not incidental engineering outcomes; they are intentionally designed **developmental intelligence nodes** where transportation, energy distribution, education systems, healthcare infrastructure, and digital networks converge into unified operational hubs.

At these nodes, HEI envisions the establishment of **multi-system integration zones**—spaces where physical infrastructure and cognitive infrastructure operate in parallel. Educational facilities supported by HDLP immersive environments, SAN-connected data relay stations, SHIELD energy stabilization systems, and community development centers all co-exist within a single infrastructural footprint. These convergence zones function as regional accelerators of development, enabling localized ecosystems of innovation, learning, and economic activity that are fully integrated into the broader national network.

A critical dimension of this infrastructure model is the **alignment of physical and digital systems through SAN integration**. Subterranean SAN corridors are envisioned to run parallel to freeway systems, creating a dual-layer infrastructure in which physical movement and data transmission are structurally synchronized. This allows real-time communication between geographic zones and digital intelligence systems, ensuring that infrastructural decisions, educational content delivery, and resource allocation are dynamically responsive to conditions on the ground.

In parallel, HDLP (Holographic Dynamic Learning Projection) systems can be deployed along freeway corridors, transforming transportation routes into **active learning environments**. Rest stops, transit hubs, and regional nodes become immersive educational zones where learners engage with context-sensitive simulations tied directly to their geographic surroundings. Agriculture regions can host real-time ecological modeling modules, coastal zones can deploy maritime systems training environments, and urban corridors can facilitate advanced systems engineering simulations. In this way, infrastructure becomes a continuous educational landscape rather than a passive transit system.

Beyond its technical and logistical functions, the decentralization of Viti Levu represents a profound **philosophical redefinition of development itself**. Traditional models of growth often rely on centralized urban cores that accumulate resources, population, and institutional power. The HEI model challenges this paradigm by distributing development evenly across geographic space, treating each region as a structurally essential component of a larger systemic whole. Growth is no longer measured by the dominance of a single metropolitan center but by the **equilibrium of opportunity across all regions**.

This distributed model reduces systemic vulnerabilities associated with overconcentration, such as congestion, infrastructure overload, resource inequality, and socio-economic fragmentation. At the same time, it fosters localized resilience by enabling each region to develop self-sustaining economic, educational, and technological ecosystems while remaining fully integrated into the national network. The result is not fragmentation, but **coherent decentralization**, where independence and interdependence coexist within a unified structural framework.

Ultimately, the North-South and East-West Freeways function as more than transportation infrastructure—they are the **physical manifestation of HEI's distributed intelligence philosophy**. They translate abstract principles of SAN connectivity, HDLP cognitive expansion, and systemic equity into tangible geographic form. Through these dual axes, Viti Levu is transformed from a centrally imbalanced geography into a living, interconnected system of continuously interacting regions.

Within this framework, geography is no longer passive. It becomes an active participant in the architecture of civilization. Roads become intelligence channels, intersections become developmental nodes, and movement itself becomes a mechanism of systemic balance. By decentralizing Viti Levu, HEI does not merely improve transportation efficiency—it fundamentally redefines how space, opportunity, and human development are organized, creating a civilization where infrastructure itself becomes a tool for achieving equilibrium, accessibility, and integrated national intelligence.

# CHAPTER 10

THE MIDWAY CIRCULAR: BUILDING THE FUTURE INTERIOR NODES

The Midway Circular represents the second and most stabilizing layer of the Viti Levu spatial reconfiguration within the HEI infrastructure model. While the North-South and East-West Freeways establish directional flow and cross-island connectivity, the Midway Circular introduces a fundamentally different architectural logic: **continuous spatial equilibrium through circular distribution**. Rather than organizing movement along linear trajectories, this system creates a closed-loop infrastructure that binds the island's interior into a unified, self-regulating network of development, exchange, and cognitive expansion.

Within the HEI framework, the Midway Circular is not simply a transportation ring. It is a **structural intelligence layer embedded into geography itself**, designed to reorganize how settlement patterns, educational access, economic activity, and energy distribution evolve across the island. By encircling the interior regions of Viti Levu, it transforms previously underutilized inland space into an active developmental field, ensuring that growth is no longer concentrated along coastal corridors but distributed evenly across the island's interior core.

At its functional level, the Midway Circular operates as a **circulatory mobility system**, enabling uninterrupted lateral movement between inland communities, agricultural zones, ecological reserves, and emerging innovation districts. Unlike linear freeway systems that optimize for point-to-point travel efficiency, the circular design prioritizes **continuous redistribution of movement and resources**. This means that population flow, goods transportation, and service access are not dependent on coastal hubs or centralized terminals but are instead facilitated through a self-contained interior loop.

This structure fundamentally reduces spatial dependency on coastal urban centers. Inland regions—historically isolated due to geography and infrastructure limitations—are no longer peripheral extensions of coastal economies. Instead, they become **integrated operational zones within a continuous feedback loop of development**. Economic activity is no longer pulled toward a single dominant axis but is instead allowed to circulate organically through multiple interior pathways. This creates a dynamic equilibrium in which no single region monopolizes access to opportunity, infrastructure, or institutional support.

A critical innovation within the Midway Circular system is the establishment of **Future Interior Nodes**, strategically positioned convergence points distributed along the circular corridor. These nodes are designed as high-density integration zones where multiple infrastructural systems intersect and reinforce one another. Each node combines transportation access, HEI educational environments, SAN data connectivity layers, SHIELD energy distribution hubs, healthcare facilities, and localized governance systems into a single cohesive spatial unit.

These nodes function as **micro-civilizational ecosystems**—self-contained environments capable of sustaining education, commerce, digital connectivity, energy stability, and community development without requiring constant reliance on external urban centers. Within each node, residents engage in continuous interaction with HEI systems, including HDLP immersive learning environments that allow real-time access to adaptive educational simulations tailored to local geography, cultural context, and economic needs. This transforms the interior of Viti Levu into a **distributed constellation of intelligent development zones**, each contributing to national stability while maintaining localized autonomy.

The integration of HEI's digital and infrastructural systems into the Midway Circular significantly enhances its systemic intelligence. SAN (Subterranean Arterial Network) corridors are embedded alongside the circular infrastructure, creating a dual-layer system in which physical mobility and digital intelligence flow in parallel. This ensures that data transfer, logistical coordination, and educational system updates are continuously synchronized across all interior regions. As a result, decision-making, resource allocation, and learning environments become dynamically responsive to real-world conditions within each node.

Simultaneously, HDLP (Holographic Dynamic Learning Projection) environments are deployed across these nodes, transforming them into **continuous learning ecosystems** rather than static infrastructure points. Within these immersive environments, learners engage with contextual simulations that reflect local agricultural systems, ecological conditions, economic activities, and cultural heritage. Education becomes embedded in geography itself, allowing knowledge to emerge directly from lived environmental interaction rather than abstract institutional separation. In this sense, the Midway Circular transforms the interior landscape into a **permanent distributed classroom system**, where learning and living are structurally unified.

Energy resilience within the Midway Circular is governed by the integration of the SHIELD system, which provides decentralized energy distribution across all interior nodes. Each node operates as a semi-autonomous energy micro-grid, capable of drawing from renewable sources such as solar, hydro, and bio-energy systems while remaining interconnected through adaptive load-balancing networks. This ensures that no single node becomes critically dependent on external supply chains, reinforcing the principle of **localized resilience within a globally coordinated system.**

The Sprinkler Protocol further enhances this resilience by enabling dynamic energy redistribution during periods of stress or disruption. In scenarios of demand spikes, environmental stress, or infrastructural strain, energy can be intelligently rerouted across nodes to maintain continuity of essential services such as healthcare, education, communication, and transportation. This creates a system in which stability is not static but continuously self-correcting, ensuring that the Midway Circular remains operational under both optimal and adverse conditions.

Beyond its technical architecture, the Midway Circular represents a profound **shift in the philosophy of spatial development**. Traditional infrastructure models are often linear in nature, prioritizing expansion outward from centralized urban cores. This results in uneven development patterns, where peripheral regions remain underdeveloped while central hubs become overburdened. The circular model fundamentally rejects this logic by introducing a **non-hierarchical spatial system**, where development is distributed evenly across all interior zones.

In this model, geography is no longer a gradient of value determined by proximity to a city center. Instead, every point along the circular network holds equivalent potential for development, innovation, and participation. This creates a system of **spatial equity**, where opportunity is not concentrated but continuously circulated through infrastructural design. The result is a balanced developmental ecosystem in which no region is permanently dominant or permanently marginal.

The Midway Circular also functions as a **feedback regulation mechanism** within the broader Viti Levu infrastructure system. While the North-South and East-West Freeways generate directional flow and expansion dynamics, the circular system absorbs, redistributes, and stabilizes these flows, preventing systemic imbalance. It acts as a corrective layer that ensures the island's developmental forces remain harmonized rather than fragmented or overly concentrated.

Ultimately, the Midway Circular and its network of Future Interior Nodes redefine the interior geography of Viti Levu as a **living,**

**intelligent, and self-regulating system**. The island is no longer understood as a set of disconnected regions linked by roads, but as a continuous spatial organism in which infrastructure, education, energy, and digital systems operate in synchronized balance.

Within this framework, development is no longer driven by external expansion or centralized accumulation. Instead, it becomes an **internal process of activation**, where every interior region is continuously engaged in cycles of learning, production, exchange, and adaptation. The Midway Circular ensures that Viti Levu evolves not through isolated centers of growth, but through a continuously interconnected interior network that sustains long-term resilience, systemic equity, and adaptive evolution across the entire island.

# FINAL CHAPTER

## THE INHERITORS OF THE EARTH

The Hologram Education Institute (HEI) is not conceived as a technological enhancement, an educational reform, or a digital innovation layer applied to existing systems. It is designed as a **foundational architecture for survival, continuity, and intergenerational legacy**. It emerges from the recognition that the modern world is increasingly defined by compounding pressures—environmental instability, resource fragmentation, demographic stress, cultural erosion, and institutional fragility. Within this context, HEI positions itself not as an optional system of advancement, but as a **structural response to civilizational vulnerability**, engineered to preserve continuity where traditional systems begin to fracture.

At its deepest level, HEI operates as an **intergenerational continuity framework**, ensuring that human knowledge, identity, culture, and infrastructure are not only preserved but continuously regenerated across time. It is not designed to function within a single generation's lifespan, but to extend across multiple cycles of human development. In this sense, HEI is not an educational model alone—it is a **civilizational operating system**, built to maintain coherence between past, present, and future states of human existence.

A central principle of this final vision is that **knowledge must remain infinite within a finite world**. While physical resources are inherently constrained, intellectual capacity is not. Through distributed systems such as HDLP (Holographic Dynamic Learning Projection) and

SAN (Subterranean Arterial Network), HEI decouples learning from physical limitation. Education becomes a continuously expanding field of accessible intelligence rather than a bounded institutional resource. Knowledge is no longer delivered through scarcity-based models of enrollment, infrastructure, or geography; instead, it exists as a **persistent, adaptive cognitive environment** available wherever human presence and need intersect.

Within this structure, SAN functions as the **substrate of continuity**, maintaining the underlying informational and infrastructural integrity of the system, while HDLP functions as the **experiential interface layer**, allowing knowledge to be embodied, simulated, and lived in real time. Together, they eliminate the traditional boundaries between learning and environment, creating a system in which education is not something accessed separately from life, but something continuously embedded within it.

Equally essential to this framework is the principle of **digital and cognitive sovereignty**. In HEI, individuals are not treated as temporary participants in institutional systems, but as permanent holders of their own intellectual and developmental identity. Every piece of data, every learning milestone, every simulation outcome, and every cultural or cognitive contribution remains under the ownership of the individual who generated it. Sovereignty is not symbolic—it is structurally enforced through distributed systems such as blockchain-based identity frameworks and SAN-linked cognitive records. In this model, identity cannot be extracted, overwritten, or centralized; it persists as an autonomous and verifiable continuity across time and systems.

This permanence of identity also extends into the **preservation of culture**, which is treated as an active, evolving system rather than a static archive. In a rapidly globalizing world, cultural knowledge is often vulnerable to dilution, displacement, or fragmentation. HEI addresses this by embedding cultural systems directly into HDLP environments, where languages, traditions, ecological knowledge, historical narratives, and social practices are not only recorded but **re-experienced in immersive form**.

Culture within HEI is not preserved as data—it is preserved as **living simulation and participatory memory**. Future generations do not simply study cultural history; they enter it, interact with it, and sustain it through engagement. A traditional practice is not reduced to documentation but reconstructed as a dynamic environment. An ecological relationship is not explained abstractly but experienced within a living model. In this way, cultural continuity becomes experiential rather than archival, ensuring that heritage remains active, evolving, and socially embedded rather than fossilized.

Education within this system is therefore designed as a **universal continuity layer for humanity itself**. It transcends geography, socioeconomic boundaries, and institutional limitations, becoming a globally distributed infrastructure of cognitive access. Every learner, regardless of location or background, becomes part of a unified but decentralized learning continuum. Access to knowledge is no longer determined by proximity to institutions or availability of physical infrastructure, but by integration into a globally connected system of distributed intelligence.

This universality reframes education as a **planetary utility**, similar in foundational importance to water, energy, or communication systems. It is continuous, adaptive, and resilient to disruption. Whether in stable or unstable conditions, the system persists, ensuring that learning remains available as a constant feature of human life rather than a conditional privilege.

Within this framework, the role of future generations is fundamentally redefined. The inheritors of this system are not passive recipients of inherited knowledge, infrastructure, or culture. They are **active custodians of systemic balance**. Their responsibility extends beyond learning into the maintenance of the systems that make learning possible. They are trained to understand complexity not only as a subject of study, but as a condition they must actively sustain.

These inheritors engage with HEI systems as **stewards of interconnected domains**—including ecological balance, technological stability, cultural continuity, and infrastructural resilience. They are not isolated specialists operating within narrow fields, but integrated participants in a larger civilizational system. Their role is to ensure that the systems supporting human development remain stable, ethical, and adaptive over time.

This introduces a new form of responsibility: **adaptive stewardship**. Rather than governing through rigid control, inheritors learn to maintain balance within dynamic systems that are constantly evolving. They are trained to recognize instability before it becomes collapse, to preserve continuity without stagnation, and to enable innovation without fragmentation. Their function is not to dominate systems, but to ensure that systems remain capable of sustaining life, knowledge, and culture across generations.

Ultimately, the inheritors of Earth do not receive a fragmented or deteriorating world in need of repair. They inherit a **deliberately constructed continuity framework**, designed to preserve complexity while enabling evolution. This framework does not eliminate change; it structures change so that it does not collapse into instability. It ensures that civilization remains adaptive without becoming chaotic, and stable without becoming rigid.

In this vision, humanity is no longer positioned at the edge of systemic breakdown, but at the threshold of a new developmental phase—one defined not by survival alone, but by **structured continuity across generations**. Knowledge, identity, infrastructure, and culture are no longer vulnerable to discontinuity; they are preserved within an integrated system designed to sustain them.

The HEI framework ultimately concludes with a single defining principle: civilization is not an accident of history that must repeatedly reinvent itself, but a **designed continuity system that can be consciously maintained**. Within this system, the future is not something that happens unpredictably. It is something that is actively shaped, preserved, and carried forward—intentionally, collectively, and intelligently—by those who inherit not only the world itself, but the responsibility to sustain its coherence.

# APPENDICES

## TECHNICAL BLUEPRINTS & THE SILENT SEEDS ARCHIVE

The appendices of the Hologram Education Institute (HEI) represent the most foundational and structurally significant layer of the entire system architecture. They are not supplementary materials, nor are they optional technical references appended to the conceptual framework. Instead, they function as the **deep operational substrate of HEI itself**—a multi-domain intelligence architecture in which every system, process, and principle defined in the main chapters is materially encoded, structurally mapped, and systemically governed.

Within this paradigm, the appendices are best understood as the **operational DNA of HEI**. Just as biological DNA determines the structure, function, and adaptive behavior of living organisms, these appendices define how HEI is constructed, how it evolves over time, and how it maintains continuity across generations, geographies, and environmental conditions. They encode not only technical specifications but also philosophical intent, ensuring that knowledge, infrastructure, ecology, and culture remain inseparably linked within a unified system of civilizational continuity.

At their highest level, the appendices establish a **multi-layered intelligence framework**, integrating physical infrastructure, digital systems, ecological preservation, and cultural memory into a single coherent architecture. Rather than existing as isolated domains, each component operates as part of a continuous feedback loop, where changes in one system propagate intelligently across all others. This ensures that HEI is not a static model, but a **self-regulating, adaptive civilization framework** capable of responding to complexity in real time.

---

## 1. SAN Architectural Diagrams — The Subterranean Nervous System of Civilization

At the core of the appendices are the SAN Architectural Diagrams, which define the **Subterranean Arterial Network as a planetary-scale cognitive infrastructure**. These diagrams do not merely illustrate connectivity between nodes; they map the underlying logic of civilization-wide information flow, treating geography itself as an integrated computational surface.

SAN is designed as a **subterranean nervous system**, where fiber-optic arteries, deep-earth routing channels, and distributed knowledge nodes form a layered communication structure beneath physical geography. These diagrams define multiple operational dimensions of the network, including:

- High-capacity data arteries linking regional intelligence hubs

- Redundant routing loops to ensure continuity under disruption

- Self-healing network protocols that dynamically reroute information

- Cognitive data channels for identity, education, and cultural transmission

- Latency-balancing grids for global synchronization

Beyond connectivity, the SAN diagrams also incorporate **geophysical integration models**, including seismic stability mapping, thermal flow distribution, and subterranean structural reinforcement systems. This ensures that the network is not only digitally resilient but physically embedded within the Earth's structural dynamics.

In its most advanced conceptualization, SAN is not a network—it is a **distributed cognitive organism**, capable of adapting, reconfiguring, and stabilizing itself in response to environmental and systemic stress.

## 2. SHIELD Energy Schematics — The Resilience Backbone of Civilization

The SHIELD Energy Schematics define the **planetary-scale energy architecture that sustains HEI's entire operational ecosystem**. Unlike traditional power grids that rely on centralized generation and distribution, SHIELD is designed as a **fully decentralized, self-balancing energy intelligence system**.

Each node within the SHIELD network functions simultaneously as:

- An energy generator
- A storage unit
- A distribution hub
- A stabilization point

These nodes are interconnected through adaptive energy routing pathways that mirror SAN's subterranean structure. This parallelism ensures that **data flow and energy flow are structurally synchronized**, forming a dual-layer infrastructure of cognition and power.

A defining feature of SHIELD is the **Sprinkler Protocol**, an adaptive redistribution mechanism that allows energy to be dynamically rerouted in response to stress conditions. During periods of high demand, environmental disruption, or infrastructural strain, the system automatically prioritizes critical sectors such as education, healthcare, communication, and ecological stabilization.

The schematics also include long-range sustainability models that simulate energy equilibrium across decades of demographic change, climate variation, and technological expansion. This ensures that SHIELD is not reactive but **predictively adaptive**, capable of maintaining balance across evolving planetary conditions.

## 3. HDLP Holographic Physics Models — The Architecture of Experiential Reality

The HDLP Holographic Physics Models define the **foundational principles behind experiential learning within HEI's immersive environments**. These models describe how information is transformed into spatial, interactive, and cognitively responsive experiences through Holographic Dynamic Learning Projection systems.

Unlike conventional digital simulations, HDLP operates as a **multi-sensory reality construction framework**, integrating:

- Volumetric rendering physics for spatial cognition

- Haptic feedback systems for embodied interaction

- Cognitive load balancing algorithms for adaptive learning

- Environmental responsiveness engines for real-time adaptation

The diagrams within this section map the transformation of abstract data into **structured experiential environments**, where knowledge is not observed but physically engaged. Learning environments dynamically reorganize themselves based on user interaction, ensuring that education becomes a responsive dialogue between learner and system.

Scalability models within HDLP ensure functionality across multiple levels of access—from personal immersive devices to large-scale shared simulation environments and full environmental overlays. This guarantees that **experiential education is not limited by infrastructure**, but expanded through adaptive system design.

## 4. Silent Seeds Biological Preservation Archive — The Memory of Life Itself

The Silent Seeds Archive represents the **ecological and genetic memory layer of the HEI system**, dedicated to the long-term preservation of biodiversity and ecological intelligence. It functions as a planetary safeguard against biological loss, environmental collapse, and ecosystem fragmentation.

Silent Seeds are not passive biological samples. They are **living informational constructs**, containing:

- Genetic material encoded with environmental context

- Growth pattern simulations

- Ecosystem interdependency maps

- Regenerative activation frameworks

This allows preserved biological systems to be not only stored but **reactivated, studied, and reintegrated** into future ecological environments if required. The archive functions as a **living biodiversity intelligence system**, preserving both species and the relational ecosystems that sustain them.

In this sense, Silent Seeds operate as a **biological continuity layer for civilization**, ensuring that ecological knowledge is never fully lost but remains recoverable through structured regeneration systems.

## 5. Fiji Long Lost History Reconstruction Data Sets — Cultural Memory Restoration Systems

The Fiji Long Lost History Reconstruction Data Sets form the **cultural and historical intelligence layer of the appendices**, focused on reconstructing fragmented or incomplete historical narratives. These datasets integrate multiple forms of recovered knowledge, including:

- Oral histories and ancestral narratives
- Linguistic evolution patterns
- Archaeological findings and site reconstructions
- Environmental and geological historical markers

Through advanced reconstruction modeling, these fragments are synthesized into **coherent cultural continuity frameworks**, enabling the restoration of historical identity even where documentation is incomplete or lost.

When integrated into HDLP environments, these datasets allow history to be experienced as **immersive cultural reality rather than static archival record**. Learners can engage directly with reconstructed environments, traditions, and historical events in fully interactive form.

This transforms cultural preservation from documentation into **active participation**, ensuring that identity is not merely remembered but continuously lived.

---

**6. Systemic Integration — The Unified Intelligence Architecture**

While each appendix component operates with distinct functionality, their true power emerges through **systemic integration**. SAN provides structural communication pathways, SHIELD ensures energy stability, HDLP enables experiential cognition, Silent Seeds preserve ecological continuity, and cultural datasets maintain historical identity.

Together, they form a **multi-domain intelligence ecosystem** in which:

- Infrastructure becomes cognitive
- Energy becomes adaptive
- Education becomes experiential
- Ecology becomes preservable intelligence
- Culture becomes immersive continuity

This integration ensures that HEI does not depend on external systems for stability. Instead, it operates as a **self-reinforcing civilizational architecture**, capable of sustaining complexity across time and disruption.

---

**Conclusion — The Appendices as the Origin of the System**

Ultimately, the Appendices: Technical Blueprints & The Silent Seeds Archive represent the **deepest expression of HEI's foundational philosophy**: that civilization must be designed as an integrated system of knowledge, infrastructure, ecology, and memory.

They are not an afterword, nor a technical appendix in the conventional sense. They are the **origin layer of the entire system**—the point at which concept becomes structure, and structure becomes continuity.

Within this framework, HEI is not described by the appendices.

**HEI is generated by them.**

# CONCLUSION

The Hologram Education Institute, as presented throughout this work, is ultimately an attempt to redefine the foundational assumptions through which civilization understands learning, continuity, and stewardship. It deliberately moves beyond the conventional framing of education as a bounded institutional service and instead repositions it as a **living civilizational substrate**—an adaptive, evolving system embedded within the operational architecture of society itself.

Within this paradigm, education is no longer a phase of life, nor a sector of governance, but a **continuous condition of human existence**. It is woven into infrastructure, environment, identity, and ecological systems, forming a unified field in which learning, memory, and adaptation are inseparable processes. Across its layered architecture— spanning immersive HDLP environments, subterranean SAN cognitive networks, SHIELD energy resilience systems, and ecological preservation archives—the HEI model consistently returns to a central principle: **civilizational survival depends not on isolated advancement, but on interconnected endurance across all systems of life**.

---

## 1. Knowledge as a Living Continuum

At its core, the system proposes a fundamental epistemological shift: knowledge is no longer treated as a finite commodity stored, guarded, or distributed through hierarchical institutions. Instead, it is understood as a **dynamic, continuously expanding field of participation**.

In this model, knowledge behaves less like a repository and more like an ecosystem—responsive, evolving, and shaped through interaction. Every learner becomes both a contributor and a node within this field, actively influencing its structure while being shaped by it in return.

This reframing dissolves inherited constraints imposed by geography, institutional exclusivity, and infrastructural inequality. In their place emerges a **distributed intelligence ecology**, where access is not a privilege of proximity but a structural feature of the system itself. Education becomes continuous not because it is made more available, but because it is **architecturally incapable of being separated from daily life**.

Resilience, in this sense, is not achieved through central control, but through **systemic dispersion, redundancy, and interdependence**.

---

## 2. Continuity as the Core Civilizational Objective

Equally significant within the HEI framework is its emphasis on continuity—not only of knowledge, but of identity, culture, ecology, and memory. This expands the definition of education beyond human cognition and into the broader domain of **civilizational persistence**.

Systems such as ecological preservation archives, biological memory repositories, and cultural reconstruction datasets are not supplementary features; they represent a deliberate effort to prevent discontinuity between past, present, and future states of human civilization.

Within this architecture, progress is no longer defined as replacement or rupture. Instead, it is understood as **continuity through transformation**—a process by which historical, cultural, and ecological knowledge is not discarded in the name of advancement, but translated into more immersive, resilient, and enduring forms.

In this sense, heritage is no longer static. It becomes **active infrastructure within future experience**, embedded within learning environments, ecological systems, and identity frameworks that allow it to remain functionally alive rather than archivally preserved.

## 3. Systemic Interdependence as Structural Intelligence

The infrastructural and technical components of HEI are not designed as independent innovations operating in parallel. Instead, they are conceived as **interdependent layers of a single civilizational operating system**.

- SAN provides cognitive and informational circulation
- SHIELD stabilizes and sustains energetic continuity
- HDLP translates knowledge into embodied experience
- Ecological archives preserve biological and environmental memory
- Cultural reconstruction systems maintain historical identity

Individually, each system addresses a distinct domain of civilization. Collectively, they form a **closed-loop resilience architecture**, where each subsystem reinforces the stability of the others.

The strength of this model does not lie in technological sophistication alone, but in **relational intelligence**—the capacity of systems to respond to one another dynamically, adaptively, and in real time. In this structure, failure in one domain does not produce collapse, but **redistributed adaptation across the system as a whole**.

Civilization, therefore, is no longer treated as fragile hierarchy but as a **self-regulating organism of interconnected functions**.

## 4. From Participation to Stewardship

Perhaps the most profound shift introduced by the HEI framework is the redefinition of the human role within civilization itself.

Individuals are no longer positioned as passive recipients of institutional output. Instead, they are reframed as **active stewards of systemic balance**—participants who are both shaped by and responsible for the systems they inhabit.

This transition transforms the meaning of agency. Responsibility is no longer limited to personal development or localized contribution, but extends into the maintenance of **social, technological, ecological, and informational equilibrium**.

In this model, progress is no longer measured solely by innovation or efficiency. It is measured by the capacity to **sustain complexity without collapse**, ensuring that advancement does not destabilize the systems upon which it depends.

Humanity's role, therefore, evolves from user to custodian—from observer to integrator within a larger civilizational framework that requires continuous care, calibration, and ethical participation.

---

## 5. The System as an Evolving Architecture

In its final articulation, HEI does not present itself as a completed or static solution to the challenges of civilization. Instead, it is intentionally designed as an **open-ended evolutionary structure**.

Its architecture assumes incompleteness as a feature rather than a flaw. It is meant to be shaped, refined, and extended by successive generations, each inheriting not only its systems but the responsibility to adapt them to new realities.

This ensures that HEI does not become obsolete through rigidity, but remains viable through **adaptive continuity**. Stability is achieved not through permanence, but through the capacity to evolve without losing coherence.

Within this framework, civilization is no longer something that is finished or achieved. It is something that is **continuously maintained, interpreted, and rebalanced across time**.

---

## 6. Final Reflection — The Beginning Embedded in the End

Ultimately, this work concludes by returning to its own origin principle: that systems of this magnitude do not end with design—they begin with responsibility.

HEI does not represent a closed theory of civilization. It represents an **infrastructure of possibility**, one that requires active participation to remain meaningful. Its completion is not found in documentation, but in stewardship.

In this sense, the conclusion is not an endpoint. It is a threshold.

It marks the transition from conceptual architecture to inherited responsibility—from design to participation—from system to continuity.

The work ends precisely where it must begin:

**with those who inherit it, and with the obligation to preserve not only what has been built, but the coherence, integrity, and intent of the vision itself.**

# AUTHOR BIOGRAPHY

Mahendra Jagir is a visionary engineer, environmental strategist, and systems thinker whose life's work is anchored in a singular and enduring question: **how can human ingenuity evolve beyond conwstruction and innovation into the deliberate design of civilizational survival?** His pursuit is not limited to the advancement of infrastructure or technology alone, but extends into the deeper architecture of stability—how societies can be structured to endure environmental uncertainty, resource volatility, and systemic disruption without collapsing into fragmentation or conflict.

At the core of Jagir's philosophy lies a fundamental redefinition of engineering itself. In his view, engineering is not simply the act of building systems that function under ideal conditions, but the discipline of designing systems that remain stable under stress. Civilization, therefore, becomes his ultimate design challenge—not as an abstract concept, but as a **living, evolving system requiring continuous maintenance, foresight, and adaptive intelligence.**

---

## 1. Cultural Memory as Engineering Intelligence

A defining influence in Jagir's worldview is his deep connection to the history, geography, and cultural traditions of Fiji and the broader Pacific region. These island societies developed sophisticated ecological knowledge systems long before the rise of modern industrial infrastructure—systems rooted in **balance, restraint, cyclical resource use, and environmental reciprocity.**

Rather than treating these traditions as historical artifacts, Jagir interprets them as **operational intelligence systems encoded in cultural memory**. He views them as evidence that sustainability is not a modern invention, but an inherited form of intelligence embedded in ancestral practice.

By integrating these indigenous ecological principles with contemporary systems engineering, climate modeling, and infrastructural design, Jagir constructs a hybrid philosophy—one that bridges **ancestral wisdom and technological foresight**. In this synthesis, tradition is not preserved as nostalgia, but activated as functional design logic for modern resilience.

---

## 2. The SHIELD Project and SAN: Dual Pillars of Resilient Civilization

Central to Jagir's vision are two interconnected systems: the **SHIELD Project** and the **Subterranean Arterial Network (SAN)**. These are not isolated technological proposals, but interdependent infrastructures forming a unified model of planetary resilience.

The SHIELD Project represents a **distributed protective energy and environmental stabilization system** designed to safeguard communities from escalating climate-related threats. Rather than relying on centralized defense mechanisms or reactive disaster response models, SHIELD is conceived as a **continuous protective layer embedded within the environment itself**. It anticipates stress conditions—such as wildfire escalation, energy instability, or extreme climate events—and redistributes protective capacity in real time to maintain equilibrium.

In parallel, the SAN system operates as a **subterranean circulatory network for civilization**, enabling the secure and uninterrupted distribution of essential resources such as water, energy, data, and logistical support. Unlike conventional infrastructure, SAN is designed with redundancy, depth, and adaptive routing logic, ensuring that no single disruption can compromise systemic integrity.

Together, SHIELD and SAN form a **dual-layer civilizational architecture**: one that protects, and one that sustains. One stabilizes external threats; the other ensures internal continuity. Their integration reflects Jagir's core belief that resilience must be both **defensive and distributive**, operating simultaneously across multiple domains of existence.

---

## 3. From Reactive Systems to Predictive Civilization Design

Jagir's work emerges from a critical observation of modern civilization: that most contemporary systems are fundamentally reactive rather than anticipatory. Infrastructure, governance, and disaster response mechanisms are often activated only after failure has already occurred.

Against this model, Jagir proposes a transition toward **predictive civilizational engineering**—systems designed not to respond to crises, but to reduce the probability of crisis emergence altogether.

In this framework, infrastructure becomes intelligent rather than inert. It monitors environmental signals, interprets systemic stress patterns, and adjusts its behavior in real time. Civilization is no longer managed through episodic intervention, but through **continuous adaptive regulation**.

This shift represents a fundamental transformation in how resilience is understood. Stability is no longer achieved through recovery after disruption, but through the **prevention of structural breakdown before it occurs**.

---

## 4. Infrastructure as a Guardian System

In Jagir's philosophy, infrastructure is not neutral. It is either passive and vulnerable, or active and protective. He advocates for a redefinition of infrastructure as a **guardian system of civilization**, capable of preserving life not only through function, but through intentional design intelligence.

Within this paradigm, roads, energy grids, water systems, communication networks, and environmental buffers are no longer isolated utilities. They become components of a **cohesive protective ecosystem**, designed to maintain equilibrium between human activity and natural forces.

This transforms infrastructure from a background support system into a **frontline stabilizing intelligence layer**, quietly operating beneath everyday life to ensure continuity, safety, and ecological balance.

---

## 5. The Shield and the Seed: A Generational Legacy

*The Shield and the Seed* represents the culmination of Jagir's intellectual and emotional journey. It is not merely a technical proposal or engineering framework, but a **multi-dimensional legacy document** that merges scientific foresight with personal responsibility.

Written with his grandchildren in mind, the work reflects a generational philosophy: that the true measure of human achievement is not what is built for the present, but what remains stable and meaningful for the future.

The "Shield" represents protection, resilience, and structural defense against uncertainty. The "Seed" represents continuity, regeneration, and the transmission of life, knowledge, and ecological balance across generations. Together, they form a dual metaphor for civilization itself: **what must be preserved, and what must be allowed to grow.**

---

## 6. Reframing Peace as Structural Intelligence

At the heart of Jagir's philosophy is a radical redefinition of peace. In his view, peace is not simply the absence of conflict, nor is it a temporary equilibrium maintained through political negotiation. Instead, peace is understood as a **condition produced by well-designed systems**.

When infrastructure is stable, when resources are distributed equitably, when ecological systems are balanced, and when communities are protected from systemic shocks, peace becomes the **natural emergent property of structural design**.

This leads to his concept of **"Shatterproof Peace"**—a form of civilizational stability that does not depend on fragility management or crisis intervention, but on the existence of **interconnected systems strong enough to absorb disruption without collapse**.

In this model, peace is no longer maintained through reaction. It is engineered through foresight.

---

## 7. Final Reflection — Engineering the Conditions for Continuity

Ultimately, Mahendra Jagir's philosophy challenges the fundamental orientation of modern development. Rather than prioritizing expansion, extraction, or reactive control, he calls for a reorientation toward **civilizational continuity engineering**—the deliberate design of systems that ensure long-term stability for both human societies and the natural world.

His vision reframes human creativity as a custodial force rather than an exploitative one. Civilization is not something to be maximized in output, but **stabilized across time**.

In this sense, his work does not simply propose new technologies. It proposes a new responsibility: that humanity must learn to design not only for progress, but for endurance.

And in doing so, it redefines the ultimate purpose of engineering itself: **not to conquer uncertainty, but to build systems capable of living within it—safely, sustainably, and in balance.**